Coding in the Cafeteria

Scratch 3.0

By Kristin Fontichiaro and Colleen van Lent

Operation Code

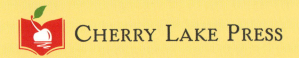

Published in the United States of America by Cherry Lake Publishing
Ann Arbor, Michigan
www.cherrylakepublishing.com

Series Adviser: Kristin Fontichiaro
Reading Adviser: Marla Conn, MS, Ed., Literacy specialist, Read-Ability, Inc.

Image Credits: ©OpenClipart-Vectors/Pixabay, 4, 6, 12, 14, 16, 18, 20; ©AnnaliseArt/Pixabay, 4, 6, 18, 20; ©AceClipart_Etsy/Pixabay, 4, 6, 18, 20; ©StarShopping/Pixabay, 4, 6, 18, 20; ©janjf93/Pixabay, 4, 6, 18, 20; Various images throughout courtesy of Scratch

Library of Congress Cataloging-in-Publication Data

Names: Fontichiaro, Kristin, author. | van Lent, Colleen, author.
Title: Coding in the cafeteria / by Kristin Fontichiaro and Colleen van Lent.
Description: Ann Arbor, Michigan : Cherry Lake Publishing, 2020. | Series: Operation code | Includes bibliographical references and index. | Audience: Grades 2-3
Identifiers: LCCN 2019035797 (print) | LCCN 2019035798 (ebook) | ISBN 9781534159303 (hardcover) | ISBN 9781534161603 (paperback) | ISBN 9781534160453 (pdf) | ISBN 9781534162754 (ebook)
Subjects: LCSH: Computer programming—Juvenile literature. | Scratch (Computer program language)—Juvenile literature.
Classification: LCC QA76.6115 .F66 2020 (print) | LCC QA76.6115 (ebook) | DDC 005.13/3—dc23
LC record available at https://lccn.loc.gov/2019035797
LC ebook record available at https://lccn.loc.gov/2019035798

Cherry Lake Publishing would like to acknowledge the work of the Partnership for 21st Century Learning, a Network of Battelle for Kids. Please visit *http://www.battelleforkids.org/networks/p21* for more information.

Printed in the United States of America
Corporate Graphics

NOTE TO READERS: Use this book to practice your Scratch 3 coding skills. If you have never used Scratch before, ask a parent, teacher, or librarian to help you set up an account at *https://scratch.mit.edu*. Read the tutorials on the website to learn how Scratch works. Then you will be ready for the activities in this book! You will practice using variables, if/then statements, copying code to other sprites, using effects to change a sprite's look, and more! Find all the starter and final programs at *https://scratch.mit.edu/users/CherryLakeCoding*.

Table of Contents

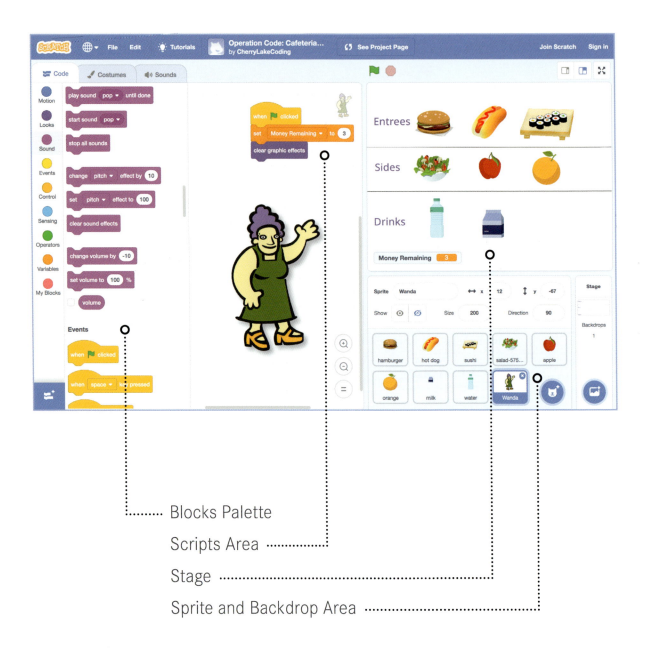

Blocks Palette

Scripts Area

Stage

Sprite and Backdrop Area

The Lunch Line Is Too Long!

I'm Ms. Wanda, head lunch lady. I need your help. The cafeteria line moves too slowly! Students take too long picking out their food. Can you help me make a game that shows what is for lunch and helps kids know what they can afford?

Pro Tip!

Scratch lets you see other people's projects and copy them to make them your own. My code contains the **sprites** and backdrop we will use in this book.

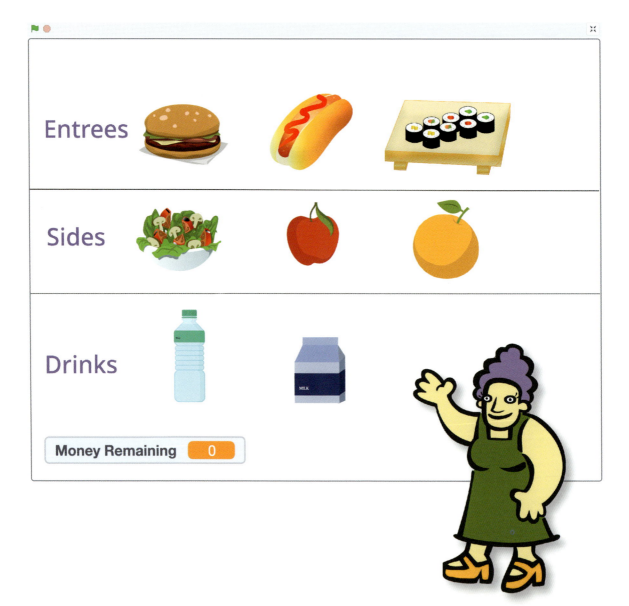

Entrees

Sides

Drinks

Money Remaining 0

What's on the Menu?

I got things started at *https://scratch.mit.edu/projects/314356341*.

Help me finish the code. We need to set prices and set the program to keep track of how much money is left.

I have lots of food for sale. So we have a lot of sprites to code!

Pro Tip!

Where am I? There's a Wanda sprite in the sprite area, but you don't see me on the stage. My sprite is hiding. You can show or hide a sprite using the eye **icons** above the sprites.

New Variable

New variable name:

Money Remaining

◉ For all sprites   ○ For this sprite only

☐ Cloud variable (stored on server)

Cancel   **OK**

Keep Track of Your Money

At school, there is probably a computer that keeps track of how much lunch money you have left. It automatically subtracts what you spend from how much you have. Our game needs to do that too.

I did this by adding a **variable** to my Wanda code. I called it Money Remaining. Click on my Wanda sprite to see the code.

Pro Tip!

To make a variable, click the Variables category and choose *Make a Variable*. Page 8 shows you how I made my variable.

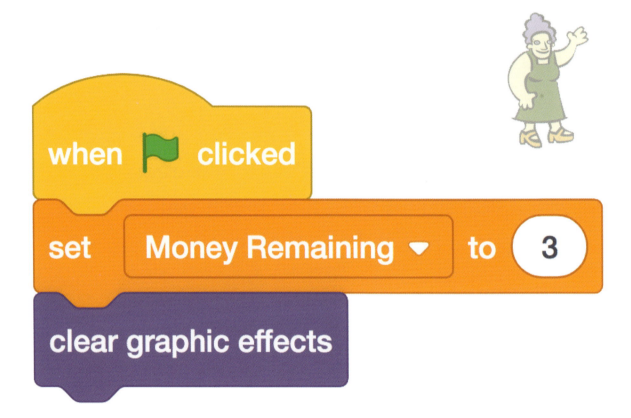

when 🚩 clicked
set Money Remaining ▼ to 3
clear graphic effects

How Much Money?

Once I created the variable Money Remaining, I needed to give it a **value** or amount. This is what the block looked like before we changed it:

Look at the code we made. You can see how we changed the block for our game. Don't worry about the  block for now.

Pro Tip!

Remember to always give your variable a **name** (like Money Remaining) and a value (like 3). If you don't update both variables, the computer can't keep track for you.

when this sprite clicked

change Money Remaining ▼ by -1

How Much Do Things Cost?

Each food item costs $1. We need Scratch to *subtract* $1 every time we click on a food item. In Scratch, when we want to subtract, we use "–" in front of the number.

Click on the Hamburger sprite. Add the code shown here.

Pro Tip!

You can find the `change my variable by ◯` block under **Variables**. Remember to change the **name** and value on the block!

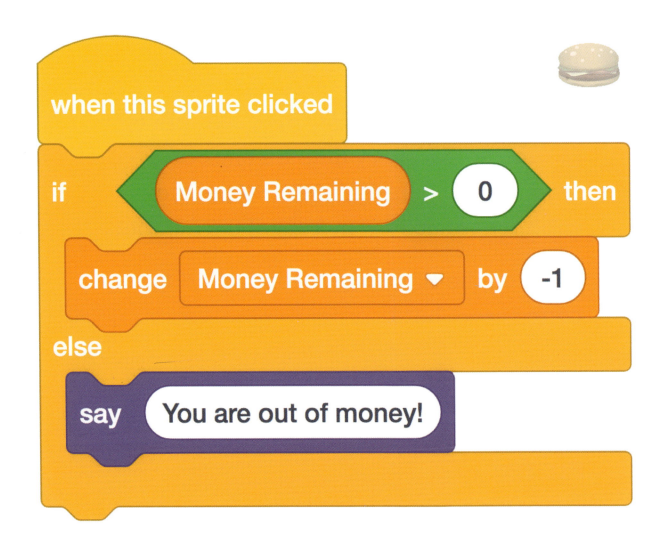

when this sprite clicked

if Money Remaining > 0 then

change Money Remaining ▾ by -1

else

say You are out of money!

Out of Money?

Make sure there is money remaining. Add the ***if ... then ... else*** Control block like I did on page 14.

This code means:

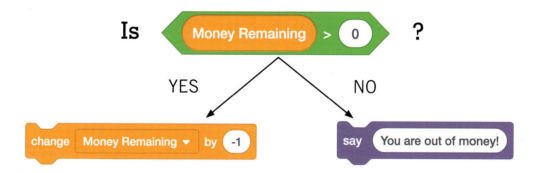

Is [Money Remaining > 0] ?

YES → change Money Remaining by -1

NO → say You are out of money!

I did on page 14.

Pro Tip!

As you buy food, your variable is going to change. The *if ... then ... else* Control block tells Scratch to make a decision during the game. It tells Scratch to decide which code to pick based on how much money is remaining at that point.

when this sprite clicked

if ⬡ Money Remaining > 0 ⬡ then

change Money Remaining ▾ by -1

change ghost ▾ effect by 30

else

say You are out of money!

Changing the Look of Food

Help me change the Hamburger sprite's color when it is clicked. We do this by adding a color effect.

Add **change color ▾ effect by 25**. Update "color" to "ghost" and "25" to "30." You can find this block under Looks.

Test your code.

Pro Tip!

Remember the **clear graphic effects** block we ignored earlier? Now it has an important job. Without it, your food would slowly fade away when you played the game over and over!

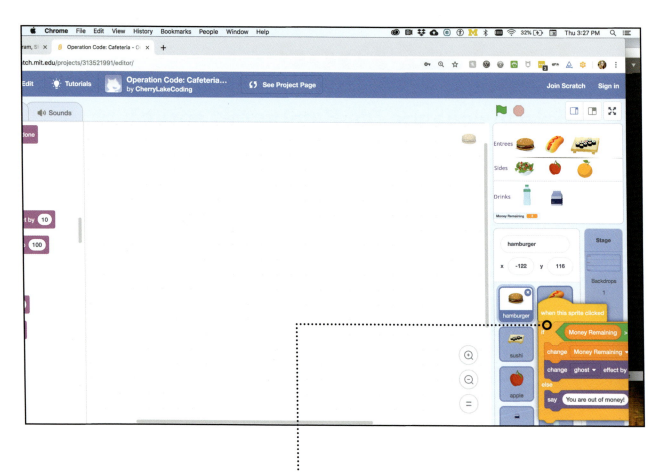

When you drag your code into another sprite, the code disappears from the scripts area. Don't worry! The code will reappear when you release your mouse.

Copying Code to Other Sprites

Time to code the other foods! We could click on each sprite and rewrite the code. But it's faster to drag the Hamburger code to each of the other sprites.

This can be tricky. Watch for the other sprites to wiggle a little. When that happens, you know you are in the right spot.

Pro Tip!

Did you copy code into a new sprite but don't see it in its script area? You may need to scroll left, right, up, or down.

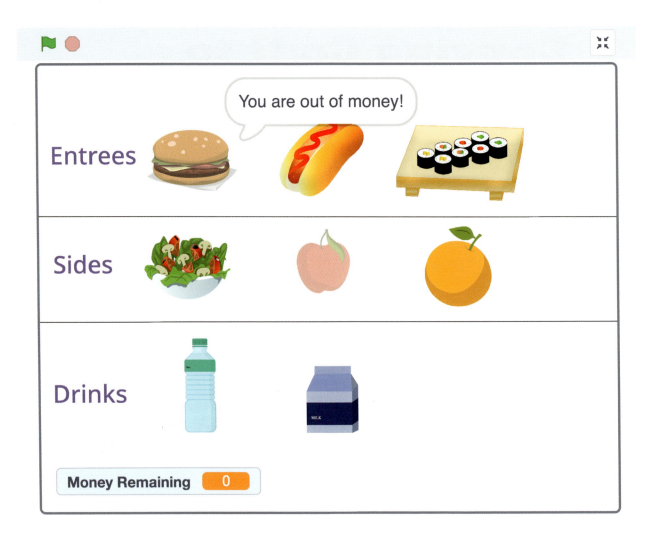

Taking Your Game to the Next Level

Thanks for helping me code. Hopefully this will make our lunch line go faster. What else could you add?

Here are some ideas:
- Add more food.
- Add sound.
- Change the color effect.
- Change the entree price to $2.

What other ideas do you have?

Pro Tip!

If you would like to see our final code, go to *https://scratch.mit.edu/projects/309591006*.

Glossary

category (KAT-uh-gor-ee) a group of things that have something in common

icons (EYE-kahnz) little pictures that you can click on

if ... then ... else (IF THEN ELS) a two-part block that says, "**IF** (this thing) is true, **THEN** (this thing) will happen, **ELSE** do (this other thing)"

name (NAME) what you call the variable when you create it; the name never changes ("money remaining" is the variable name in this book)

sprites (SPRYTS) characters or objects in Scratch

value (VAL-yoo) the information stored by the computer; the value can change many times or always stay the same (Money is the value in this book)

variable (VAIR-ee-uh-buhl) a changeable amount recorded in Scratch's memory

Find Out More

Books

LEAD Project. *Super Scratch Programming Adventure!* San Francisco, California: No Starch Press, 2019.

Lovett, Amber. *Coding with Blockly.* Ann Arbor, Michigan: Cherry Lake Publishing, 2017.

Websites

Scratch

https://scratch.mit.edu

Get started with Scratch at this website.

Scratch Wiki: Variable

https://en.scratch-wiki.info/wiki/Variable

Learn more about variables from the Scratch team.

Index

About the Authors

Kristin Fontichiaro teaches at the University of Michigan School of Information. She likes working with kids on creative projects from coding to sewing to junk box inventions. She has written or edited almost 100 books for kids.

Colleen van Lent teaches coding and Web design at the University of Michigan School of Information. She has three cool kids and a dog named Bacon. She wishes she could touch her toes.